# YOUR KNOWLEDGE HAS VALUE

# The Separation of Uranium from Seawater using Amidoxime containing Polymer

Sylvester Gyasi

**Bibliographic information published by the German National Library:**

The German National Library lists this publication in the National Bibliography; detailed bibliographic data are available on the Internet at http://dnb.dnb.de.

ISBN: 9783346865564
This book is also available as an ebook.

## Disclaimer

The ideas and design of this project should not be used without first assessing and consulting a professional because it based on theoretical information gathered from literature sources. Our thoughts and ideas could change from time to time as we learn and acquire more knowledge in this field. It should be noted that the conclusions of these design project are highly hypothetical sequence of processes and are based on unproven extrapolations and untried technology.

**ABSTRACT**

Technically, it is possible to separate uranium from seawater. This design paper seeks to propose a process intensified way to separate uranium from seawater using amidoxime containing polymer due to the high affinity of the uranyl–carbonate complex towards the polyamidoxime resin. The process justification was based on the economic criteria: production cost, physicochemical criteria: concentration and reactivity, technical criteria: extracting methods from a complex aqueous system. The separation process firstly is the adsorption of uranyl ions using ionic exchange resin in the form of a chelating polymer adsorbent. A buoyant platform with adsorbent polymer made of polyethylene fibres that comes into contact directly with sea water powered by a wind turbine which is mounted on top of the platform, continuously takes the adsorbent from the seawater after campaign time through an elution and regeneration process and then returns it to the seawater for reuse. After the adsorbent has been loaded with uranyl ions, additional downstream processes is explored to recover the uranium including cleansing of the adsorption polymer to remove organic materials, desorption: elution of the adsorbed uranium ions from the adsorption polymer with a suitable solution, purification of the eluent: removal of other desorbed compounds, concentrating the solution, solvent extraction of uranium from the solution with a mixture of an organic solvent and a specific complexing agent and finally conversion into yellow cake or uranium oxide $U_3O_8$. Next, factors that goes into the Uranium separation cost and financial justifications have been included. In future, we recommend researchers to make use of organically sourced, biodegradable substrates such as chitin which could be a good source of green chemistry alternative to conventional plastic polymers that represents a potential source of an uneasy degradable solid pollutant.

**Keywords**: Uranium extraction Uranyl ion; Seawater; Adsorption polymer; Adsorption kinetics; Separation; Reusability; Wind turbine; buoyant platform; Chelating ligand.

# Table of Contents

List of Tables

List of Figures

**Figure 3:** An image of a three-dimensional view of continuous uranium separation system

with adsorbent belt looped around the turbine mast proposed by Picard et al. (2014).

Adapted from "Extraction of Uranium from Seawater: Design and Testing of a Symbiotic

## INTRODUCTION

### Overview

Dissolved uranium content in seawater is estimated to be 4.5 billion metric tons. It is approximately 3.3 ppb, and its concentration varies in direct proportion to changes in salinity. Since seawater is slightly basic under normal coastal conditions (pH 8.0±0.4) uranium exists primarily as $UO_2(CO_3)_3^{-}$ (Tamada, 2009). Currently, 61000 metric tonnes of uranium are needed annually to meet global energy demand (Tamada, 2009). Since it represents an essentially unlimited supply, this sustainable source if convincingly shown to be recoverable at a suitable cost, environmentally friendly and energy efficient way, can establish a huge alternative to conventional methods for example, one (1) gram of uranium-235 can theoretically produce through nuclear fission, as much energy as burning 1.5 million grams of coal (Einsley, 2001).

### Aim and Objectives

To propose a process intensified design project for separation of uranium from seawater.

To make recommendations for further design improvement for future research works.

### LITERATURE REVIEW

### Uranium Ion Bonding Mechanism

The dominant form of the uranyl ion in seawater is the tricarbonate complex, $UO_2(CO_3)_3^{-}$ which associates with $Ca^{2+}$, a major cation in seawater, to form a $Ca_2UO_2(CO_3)_3$ (aq) ternary complex, the dominant uranyl species in seawater. The uranyl ion binds to the two adjacent amidoxime ligands on the adsorbent material to form a chelate complex.

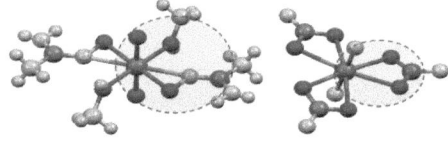

**Figure 1**

**Figure 2**

**Figure 1:** An image of how the amidoxime ligands binds with the Uranyl cation to form a chelate complex. Adapted from "How Amidoximate Binds the Uranyl Cation"by Vukovic, S.; Watson, L. A; Kang, S. O.; Custelcean, R.; Hay, B. P. Inorganic Chemistry (2012), 51, page 3855-3859.

**Figure 2:** An image of the structure of $UO_2(CO_3)_3^{4-}$ complex when it is bonded and how it exists in the seawater. Adapted from "Structure of,$UO_2(CO_3)_3^{4-}$complex" by H. Sodaye et al (2009), page 9–32.

## Alternative Methods of Separation

Several methods for separation of uranium on large scale have been explored including solvent extraction, ion exchange, foam separation, co-precipitation, biological separation and adsorption (Sodaye et al, 2007). Adsorption method using amidoxime containing polymer drew attention due to the high affinity of the uranyl–carbonate complex towards the polyamidoxime resin.

## Separation Method

Selective adsorption of $(UO_2)^{2+}$ via chelating polymer adsorbent by Kim et al (2013) has proven to be a promising field and interest to most researchers due to cost, adsorption capacity and environmental footprints (Zhang et al 2013; Seko et al 2003 and Anivudhan et al 2011). Membrane filtration, coagulation and precipitation were found to have issues as high operating cost, durability and toxicity (Kanno et al 1984; Van Reis et al 2007).

**Constraints**

The extremely low concentration of uranium and the relatively high concentrations of other number of dissolved chemical species in seawater have important consequences on the adsorption kinetics that would be able to separate uranium from seawater. For example, some ions particularly vanadium and iron are relatively adsorbed in large whilst amounts of calcium and magnesium are adsorbed in small amounts. This may necessitate a complicated elution process (Sodaye et al, 2007).

**ANALYSIS SECTION**

**Process Selection and Description**

Selective adsorption via chelating polymer will be used in the extraction of uranium from seawater. A platform with buoyant braid adsorbent made of polyethylene fibres on a polypropylene trunk that has direct contact with sea water powered by a floating wind turbine which continuously takes the adsorbent from the sea after campaign time through an elution and regeneration process and then returns it to the sea for reuse. The elution, regeneration and purification plant are housed on the upper platform out of the seawater.

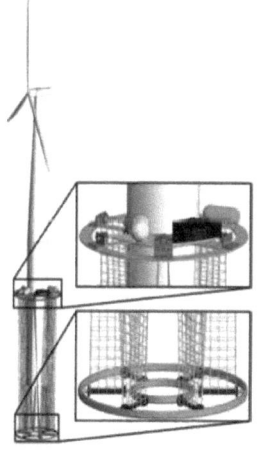

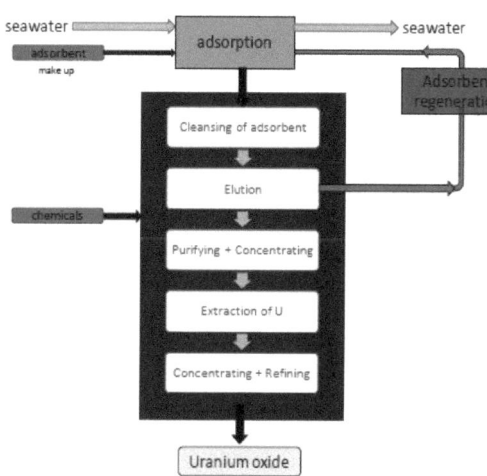

**Figure 3**                    **Figure 4**

**Figure 3:** An image of a three-dimensional view of continuous uranium separation system with adsorbent belt looped around the turbine mast proposed by Picard et al. (2014). Adapted from "Extraction of Uranium from Seawater: Design and Testing of a Symbiotic System" Nuclear Technology, 188(2), 2014 by M. Picard, C. Baelden, Y. Wu, L. Chang, and A. H. Slocum.

**Figure 4**: A image of the Process flow chart of the separation process of uranium from seawater using amidoxime adsorbents designed for the purpose of this work.

## Process Selection Justification

The uranium separation process consists of three basic steps: adsorbent synthesis, adsorbent deployment, and elution and purification of uranium. The adsorbent material is reusable, though not indefinitely, and it undergoes an autonomous process which consist of multiple deployment, elution and purification cycles. Also, to forgo the cost of active pumping it was important to make use of ocean current to flow past the adsorbent system. Hence, eliminating the offshore mooring and labour cost; reduces size of area of land required; increasing the energy output; high/maximum recovery of uranium; reduction in capital and operating cost.

## Equipment Design and Specification

Individual equipment design that is required for the design project consist of a wind turbine and buoyant platform, an adsorbent synthesis design, and elution and regeneration tanks.


### Wind Turbine and Buoyant Platform

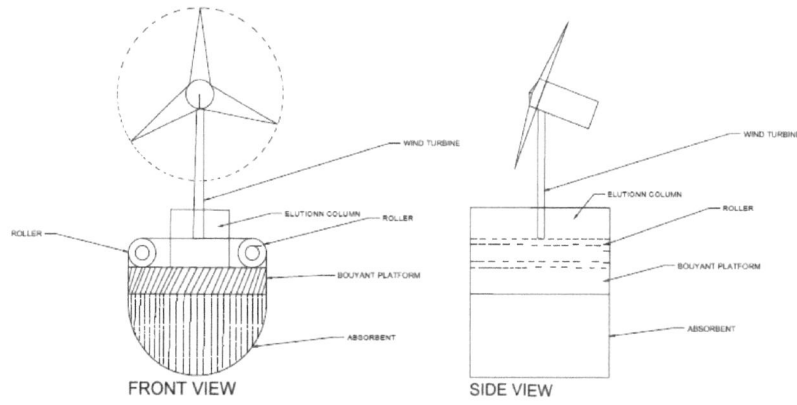

**Figure 5**

**Figure 5:** An image of the front view and side view equipment design of wind turbine and

buoyant platform designed for the purpose of this work.

## Amidoxime polymer synthesis

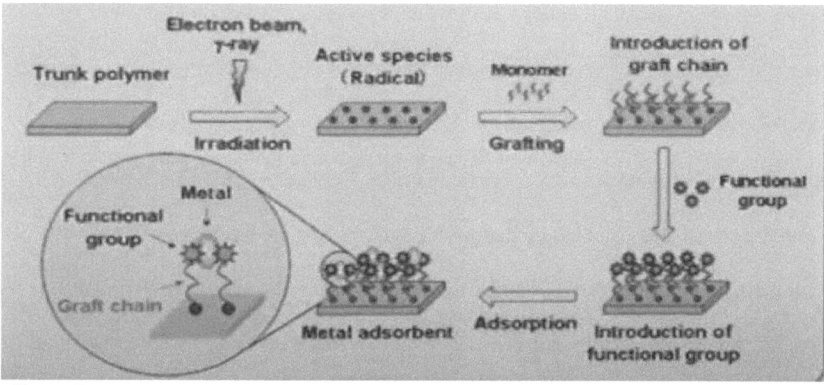

**Figure 6**

**Figure 6:** An image of how a trunk polymer is synthesised to form an amidoxime polymer

used as an ion exchanger resin for the separation of uranium. Adapted from "Current Status

of Uranium Collection from Seawater" presented at Lawrence Berkeley National Laboratory,

by Tamada, M et al., (2010).

## Elution and Regeneration Flow Chart

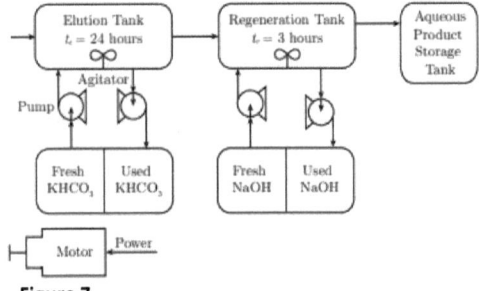

**Figure 7**

**Figure 7:** An image of the flow chart of elution and regeneration process using KHCO3 and

NaOH treatment for the process. This flow chart was designed for the purpose of this work.

### Elution and Regeneration Overview

Because uranium is strongly bonded to amidoxime groups in the adsorbent, its removal

from the adsorbent material without causing damaging effects on adsorbent performance is

a challenging problem, for example using hydrochloric acid causes a significant reduction to

the adsorbent's loading capacity for repeated usage. The elution of uranium in the elution

process is to use potassium bicarbonate with addition of sodium bicarbonate to prevent

hydrolysis. Moreover, after the acid elution, the adsorbent requires a NaOH reconditioning

process, which involves heating the adsorbent in 2.5% NaOH solution at 80°C to regenerate

the active functional groups for reuse. The NaOH is vital for the removal of seawater organic

matter that are adsorbed on the chelating polymer.

### Heat and Mass Requirement

Based on the general conservation law: **Output= Input + Generation – Consumption –**

**Accumulation**. For steady state, accumulation for mass and energy balances will be zero (0).

For mass balance, there will be no generation and consumption. Hence, the final equation reduces to: **Output =Input+ Generation -Consumption**.  Active pumping would require high energy since it uses high power hence it would also be an energy input. Energy costs for plant operation and maintenance offshore consumes large amounts of electricity. A very energy-intensive process is the regeneration process of the eluant by steam stripping because it is energy consuming.

### Plant Availability and Location

A suitable plant location will depend on these factors, seawater currents, seawater surface winds, seawater surface temperature, wind profile exponent, turbulence intensity. The plant capacity is set at 2000 tonnes per year and will operate 24 hour per day without any downtimes.

### Environmental Impact

With the current uranium adsorbent materials, Uranium and vanadium are the predominant metal adsorbed. This could be a potential concern since some marine inhabitants utilize vanadium, but there is little evidence that organisms require vanadium or uranium as an essential element. The utilization of uranium seems to be a subsidiary need for the marine organism and not essential for life. In contrast to uranium, vanadium, appears to have a much larger function in biological systems.

### Pipework and Pumping Systems

**Pipework**: In specifying the pipework for the process design, the parameters considered are material of construction, optimum diameter, schedule number, nominal size, internal diameters and pipe wall thickness. The preferred schedule number is 80 constructed from 304 stainless steel.

**Pumps:** In specifying the pumping systems for the process design, the parameters considered are material of construction, pump type, power of pump, type of fluid transported and pump head. The preferred pump is a Centrifugal pump that is constructed from a stainless steel.

**Health and Safety**

**Table 1 Health and Safety**

| Material Hazards | Process Hazards | Mitigation |
|---|---|---|
| 1. KHCO3, NaOH and HCL are highly corrosive materials | 1. Hazards from Equipment and instrument failure | 1. Pressure relief valves |
| 2. Metallic materials may be good electrical conductivity. | 2. Pressure and temperature deviation | 2. Insulation of hot surfaces |
| 3. Materials of construction can be hazardous due to sharp edges and hot surfaces. | 3. Loss of containments | 3. Periodic assessment of working condition of equipment. |
| 4. Materials for elution and regeneration are toxic | 4. Fire and Ignition from electrical equipment. | 4. Drainage or spillage containment barriers |
| | 5. Noise and visual pollution | 5. Immediate replacement of faulty equipment part |

## Safety Protection Devices

### Table 2 Safety Protection Devices

| Device | Function |
|---|---|
| Electrolytic system | Prevent marine organisms from settling down and multiplying on the surface of the pipes. |
| Ultrasonic system | High frequency waves to reduce biofouling. |
| Electronic monitoring device | To monitor overall process system and environment. |

## Control Strategies and Systems

### Table 3 Control Strategies and System

| Process Variable | Control system | Process unit/equipment |
|---|---|---|
| Temperature | Thermocouples | Elution and regeneration tanks |
| Pressure | Pressure control valves | Elution and regeneration tanks |
| Flow rate | Flow meters | Storage tanks, Elution and regeneration tanks |
| Fluid level | Level sensors | Storage tanks, Elution and regeneration tanks |
| pH | pH sensors | Elution and regeneration tanks |
| Composition | Composition/ Electrochemical analyser | Elution and regeneration tanks |
| Safety | Alarms and trips | Turbine, Elution and regeneration tanks |

## Capital Cost Estimations and Financial Justification

The adsorption synthesis, adsorbent capacity, recycle efficiency of the adsorbent, chemicals usage, deployment mechanism, plant operation and equipment cost have the direct relevance to the cost of the uranium separation from seawater.

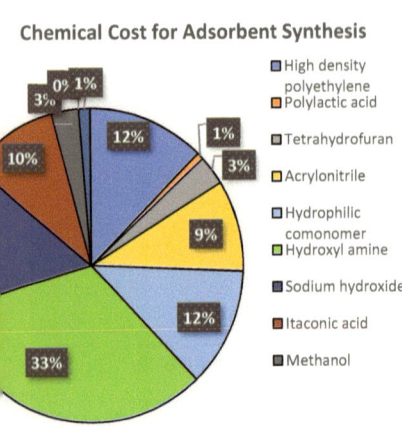

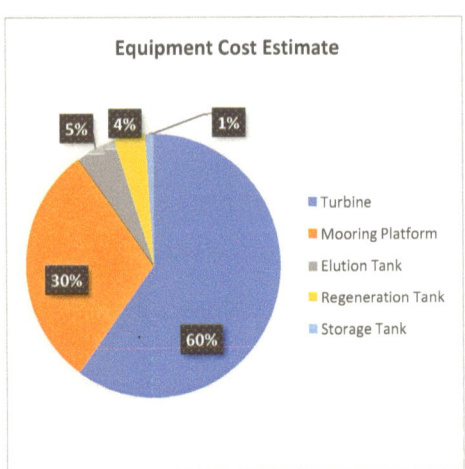

**Figure 8**

**Figure 9**

**Figure 8:** An image of a graph for the estimation of $555/kg of uranium separation cost using estimated figures for the purpose of this work.

**Figure 9:** An image of a graph for the estimation of equipment cost using estimated figures for the purpose of this work.

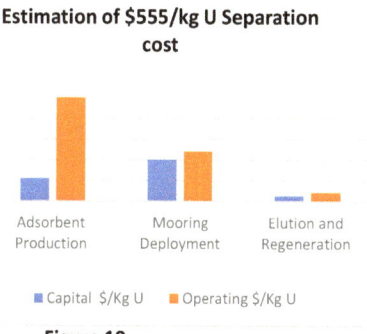

Figure 10

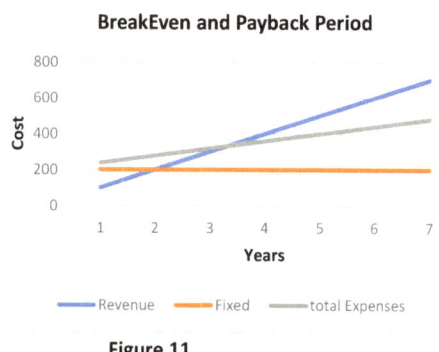

Figure 11

**Figure 10:** An image of a graph for the chemical cost of adsorbent synthesis using estimated figures for the purpose of this work.

**Figure 11:** An image of a graph for the breakeven and payback period analysis using estimated figures for the purpose of this work.

## RECOMMENDATION

It is recommended that adsorbent synthesis should be made of organically sourced, biodegradable substrates such as chitin which could be a good source of green chemistry alternative to conventional plastic polymers that represents a potential source of an uneasy degradable solid pollutant.

## CONCLUSION

Although there are several ways for uranium separation, the offshore wind turbine( to produce energy that would be required) on a buoyant platform that continuously takes the adsorbent designed to have a high affinity of the uranyl–carbonate complex from the seawater after campaign time through an elution and regeneration process( housed on the platform above the seawater) and then returns the adsorbent back to the seawater for the seawater current to flow past the system( to forgo the high energy demand for active pumping) for reuse for some number of cycles before degrading( to save cost implication of adsorbent production) has proved to be the most process intensified method for separation of uranium from seawater. The cost analysis and payback period on investment has been financially justified.

**REFERENCES**

1. M. F. Byers and E. Schneider. Uranium from Seawater Cost Analysis: Recent Updates.2015.

2. M. F. Byers, M. N. Haji, A. H. Slocum, and E. Schneider. A Higher Fidelity Cost Analysis of Wind and Uranium from Seawater Acquisition symBiotic Infrastructure. 2016.

3. J. Kim, C. Tsouris, R. T. Mayes, Y. Oyola, T. Saito, C. J. Janke, S. Dai, E. Schneider, and D. Sachde. Recovery of Uranium from Seawater: A Review of Current Status and Future Research Needs. Separation Science and Technology, 48:367–387, 2013.

4. M. Tamada. Current status of technology for collection of uranium from seawater. Japan Atomic Energy Agency,2009.

5. A. Zhang, T. Asakura, and G. Uchiyama. The adsorption mechanism of uranium (VI) from seawater on a macroporous fibrous polymeric adsorbent containing amidoxime chelating functional group. Reactive and Functional Polymers, 57(1):6776, 2003.

6. Z. Xing, J. Hu, M. Wang, W. Zhang, S. Li, Q. Gao, and G. Wu. Properties and evaluation of amidoxime-based UHMWPE fibrous adsorbent for extraction of uranium from seawater. Science China Chemistry, 56(11):1504–1509, 2013.

7. G. A. Tularam and M. Ilahee. Environmental concerns of desalinating seawater using reverse osmosis. Journal of Environmental Monitoring, 9(8):805–813, 2007.

8. University of Idaho. Uranium Recovery from Seawater: A Nation-Wide Consortium

for Sustainable Energy - University of Idaho. http://uraniumfromseawater.

engr.utexas.edu/partners/university-idaho, Accessed: 2017-04-15.

9. M. Tamada, N. Seko, N. Kasai, , and T. Shimizu. Cost estimation of uranium recovery

from seawater with system of braid type adsorbent. Transactions of the Atomic Energy

Society of Japan, 5(4):358–363, 2006.

10. T. Sugo, M. Tamada, T. Seguchi, T. Shimizu, M. Uotani, and R. Kashima. Recovery

System for Uranium from Seawater with Fibrous Adsorbent and Its Preliminary Cost

Estimation. Journal of the Atomic Energy Society of Japan, 43(10):1010–1016, 2001.

11. N. Seko, A. Katakai, M. Tamada, and S. Takanobu. Fine Fibrous Amidoxime Adsorbent

Synthesized by Grafting and Uranium Adsorption Elution Cyclic Test with Seawater.

Separation Science and Technology, 39(16):3753–3767, 2004.

12. M. Schwartz, D. Heimiller, S. Haymes, and W. Musial. Assessment of Offshore Wind

Energy Resources for the United States Assessment of Offshore Wind Energy Resources

for the United States. Technical report, National Renewable Energy Laboratory, 2010.

13. E. Schneider and D. Sachde. The Cost of Recovering Uranium from Seawater by a Braided

Polymer Adsorbent System. Sci. Glob. Sec., 21(2):134–163, 2013.

14. D. J. Sachde. Uranium Extraction from Seawater: An Assessment of Cost, Uncertainty

and Policy Implications. Master's thesis, The University of Texas at Austin, 2011.

15. M. Picard, C. Baelden, Y. Wu, L. Chang, and A. H. Slocum. Extraction of Uranium from

Seawater: Design and Testing of a Symbiotic System. Nuclear Technology, 188(2), 2014.

16. J. Park, G. A. Gill, J. E. Strivens, L.-J. Kuo, R. Jeters, A. Avila, J. Wood, N. J. Schlafer, C. J.

Janke, E. A. Miller, M. Thomas, R. S. Addleman, and G. Bonheyo. Effect Of Biofouling on The

Performance of Amidoxime-Based Polymeric Uranium Adsorbents. Industrial & Engineering

Chemistry Research, (15), 2016.

17. P. Hay, C. M. Wai L.J. Kuo, G. Gill, G. Tian. Bicarbonate Elution of Uranium from Amidoxime-Based Polymer Adsorbents for Sequestering Uranium from Seawater, 2017.